TAX REFORM

Dados Internacionais de Catalogação na Publicação (CIP)
(Câmara Brasileira do Livro, SP, Brasil)

```
Coletta, Osvaldo Dalla
   Tax reform : current tax system / Osvaldo
Dalla Coletta. -- 1. ed. -- Santo André, SP :
Ed. do Autor, 2022.

   ISBN 978-65-00-56571-3

   1. Administração pública 2. Direito tributário
3. Finanças públicas 4. Imposto - Brasil - Legislação
5. Reforma tributária 6. Sistema Tributário Nacional
(Brasil) I. Título.
```

22-135629   CDU-34:336.2(81)

Índices para catálogo sistemático:

1. Brasil : Reforma tributária : Direito tributário
       34:336.2(81)

Aline Graziele Benitez - Bibliotecária - CRB-1/3129

Graphic design and Desktop publishing:

Eliane Otani/Visão Editorial

Osvaldo Dalla Coletta

Economist CRE – SP n. 6.813

TAX REFORM

Current Tax System

CONTENTS

- **Tribute:** Tributes are made up of taxes, fees and contributions.

- **Contributors:** only individuals are tribute payers, because they are the final consumers who buy and pay the prices of products and services, prices that include taxes. Legal entities (companies) are not final consumers and are not tribute payers because they do not pay tribute, they only collect tribute that are included in the sales price of products and services, which are purchased and paid by individuals, who are end consumers.

- **Tributary progressiveness:** tributary progressivity taxes in proportion to each tribute payer's ability to pay. In other words, it tributes proportionally to the income of each tribute payer or charges more from those who earn more and charges less from those who earn less.

- **Dumping:** it is an unfair business prac-tice. For example, a certain imported prod-uct arrives in Brazil at a price lower than the cost of the product in the exporting country. In this case, the import tribute is charged, so that the cost of that imported product is equal to the cost of the product in the export-ing country.

- **Algorithm:** It is a logical sequence of instructions to produce a certain result. For example, to calculate 2 x 3 = 6, the computer programmer writes a logical sequence of instructions, equivalent to what is done with a manual calculator when pressing the keys (2), (x), (3) and (=), to obtain the result 6.

FEDERAL TAX REFORM

The federal tax reform will replace all current federal taxes with a progressive national contribution (PNC), which will be levied on all individuals residing in Brazil.

There is only one exception: the import tribute, which must be maintained to preserve its anti-dumping regulatory function.

The PNC calculation basis will be the total monthly gross income of each individual minus the amount equivalent to one minimum wage.

The collection of this PNC will be done automatically by the Federal Revenue's electronic processing system, through withholding at source, at the time of payments made to each individual. For each individual and up to the fifth business day of the current month, the algorithm of the electronic processing system will add up all the amounts withheld in the month prior to the current month and, from this sum, *WILL DECREASE* the value corresponding to a minimum wage, in order to obtain the PNC calculation basis and with this calculation basis, select the rate from the progressive table of the income range corresponding to the value of the obtained calculation basis, to calculate the PNC value due in the month prior to the current month. When there is a difference between the total monthly amount withheld in the month prior to the current month and the total monthly amount due in the month prior to the current month, the amount of the difference will be debited or credited to the individual's CPF.

The progressive national contribution (PNC) will be credited automatically and directly to the Federal Revenue's bank account and will be charged according to the following progressive table:

Minimum wages	Rates (%)
0 to 1	1
1 to 2	2
2 to 3	3
3 to 4	4
4 to 5	5
5 to 6	6
6 to 7	7
7 to 8	8
8 to 9	9
9 to 10	10
10 to 11	11
11 to 12	12
12 to 13	13
13 to 14	14
14 to 15	15
15 to 16	16
16 to 17	17
17 to 18	18
18 to 19	19

Minimum wages	Rates (%)
19 to 20	20
20 to 21	21
21 to 22	22
22 to 23	23
23 to 24	24
24 to 25	25
25 to 26	26
26 to 27	27
27 to 28	28
28+	29

Progressive taxation taxes proportionally to the contribution capacity of each taxpayer. In other words, charge less from those who earn less and charge more from those who earn more. For example, a person who earns BRL 1,300.00 per month will pay 1% of PNC, which is equal to R$ 13.00, and will keep R$ 1,287.00 for him. A person who earns BRL 35,000.00 per month will pay 29% of PNC, which is equal to

BRL 10,150.00, and will keep BRL 24,850.00 for them.

The algorithm of the electronic processing system will make the daily control of the collection, in order to collect the amount contained in the Federal Government budget, neither more nor less, and, if necessary, it will make daily adjustments to the rates of the progressive table, increasing or decreasing rates, starting with +0.01% or −0.01%, so that the amount collected is equal to the amount set out in the Federal Government budget, regardless of upward or downward variations in economic activity.

The electronic records of payments made to individuals will be made in real time (online), directly on the website of the Federal Revenue Service. These electronic records will contain the total amount of the payment, the amount of the PNC withheld, the CNPJ or CPF of the payer, the CPF of the recipient and a code with the other information needed by the Federal Revenue and the Brazilian Institute of Geography and Statistics (BIGS).

When payments to individuals cannot be recorded in real time (online), payers will electronically record these payments at the end of each day or at the end of each week

This tax reform ensures that the Federal Government has its collection (revenue) preserved, after its approval and implementation.

After the approval and implementation of this tax reform, the Federal Government will be prohibited from: collecting more than what is stated in the annual budget, spending more than it collects, borrowing money, lending money and being guarantor of borrowers, as well as having to pay off all its debts already contracted within a maximum period of twenty years.

When tributary payers want to reduce the tribute burden, they will need to interact with their representatives in Congress, so that the budget for the following year is approved in accordance with the tribute burden reduction intended by tributary payers.

Considering that the Federal Government will neither be able to collect more nor spend more than what is foreseen in the annual budget, it will be obliged to make a financial reserve fund destined to the expenses necessary to meet the needs arising from possible natural disasters and health emergencies, and also, to take out a subsidiary liability insurance policy for each public work carried out, to ensure the good quality of the public tender notice, the work project, the quality of the construction of the work, compliance with the deadline for delivery of the work, as well as to guarantee the payment of eventual indemnities for damages caused against third parties until the end of the warranty period of each contracted work.

It will be in the insurer's interest to act to avoid failures in the public tender notice, project failures, failures in the construction of the work, delay in the delivery of each insured public work and possible damages caused against third parties.

Without prejudice to the responsibility of the construction company's technical engineer responsible for each public work contracted by the Federal Government, the insurance company will have a subsidiary technical responsible engineer of its own for each insured public work, as a preventive measure to avoid the possibility of having to pay a fine for delay in delivery of the assured work and financially assume the losses resulting from failures in the public tender notice, in the design and construction of the work, for the repairs and repairs of construction defects and any damages caused against third parties until the end of the warranty period of each public work assured.

Without prejudice to the responsibility of the engineer in charge of the construction company's technical department, the engineer of the contracted insurance company and subsidiary technical manager of the insured public work will have the following rights: free access to the public work assured until the end of the construction and delivery of the public work assured. The participation of the subsidiary technical engineer of the contracted insurance company during the preparation of the public tender notice is important to prevent possible technical failures, such as, for example, guaranteeing the correct technical specification of the macadam for the paving of roads, avenues and streets, in other words, ensure that the technical specification of the macadam is adequate to withstand the traffic intensity of vehicles with their respective maximum loads (the weight of each vehicle plus the maximum load they can carry), to prevent the asphalt surface from showing deformations and holes shortly after paving.

After the approval and implementation of this federal tax reform and state, district and municipal tax reforms similar to the federal tributary reform, legal entities (companies) will only make the monthly deposit of the Severance Indemnity Fund (SIF) of each worker and, when applicable, they will pay collect import tribute. Payments for services and payments for labor, production, sale and transport of machinery, products and goods will be completely free from taxation and completely free from government bureaucratic embarrassment.

STATE TAX REFORM, DISTRICT AND MUNICIPAL

The tax reform for all states, for the Federal District and for all municipalities is the same as the federal tax reform. In the explanatory text of the state, district and municipal tax reform, in comparison with the explanatory text of the federal tax reform, what changes are only the words Federal, federal, Brazil, National Congress, and the acronym PNC (progressive national contribution), replacing them respectively by the words State or District or Municipal, state or district or municipal, State or Federal District or Municipality, State Assembly or District Assembly or City Council and the acronym PNC by PSC (progressive state contribution) or DCP (district contribution progressive) or PMC (progressive municipal contribution).

The state, district and municipal tax reforms will replace all current state, district and municipal taxes with their respective progressive state contribution (PSC), progressive district contribution (PDC) and progressive municipal contribution (PMC), which will be collected from individuals residing in their respective territories, state, district and municipal, through their respective electronic sites and automatically and directly credited to the bank accounts of the respective state, district and municipal revenues.

These tax reforms, after their approval and implementation, ensure that all states, the federal district and all municipalities have their collections (revenues) preserved and ensure the fiscal autonomy of the states, the federal district and the municipalities, with the elimination of the current fiscal dependency caused by current federal revenue transfers to the states and the

federal district, and from the states to the municipalities.

When taxpayers want to reduce the tax burden, they will need to interact with their legisla-
tive representatives, so that the budget for the
following year is approved in accordance with
the reduction in the tax burden intended by taxpayers.

After the approval and implementation of this
tax reform, state, district and municipal governments will be prohibited from: collecting more
than is included in the respective annual bud-
gets, spending more than they collect, borrowing money, lending money and being guarantors
of money borrowers borrowed, as well as having
to settle all their debts already contracted within
a maximum period of twenty years. These same
governments will be obliged to set up a financial
reserve fund for the necessary expenses with

meeting the needs arising from possible natural disasters and health emergencies and will also be required to take out a subsidiary insurance for each public work that carry out, the same as that contained in the reform federal tax.

CURRENT TAX SYSTEM

Concepts

- **Taxes:** taxes are made up of taxes, fees and contributions.

- **Examples:** the tax on consumption of goods and services (TCGS) is a tax; the boarding tax is a fee; and the contribution to the National Institute of Social Security (NISS) it is a contribution.

- **Types of taxes:** direct and indirect.

- **Direct taxes:** direct taxes, such as the tax on urban territorial property (TUTP) and the tax on income and earnings of any nature (IR), are characterized by being personalized, since they are charged, respectively,

according to value of each person's property
and each person's income.

- **Indirect taxes:** indirect taxes are characterized by being impersonal and regressive; the percentage rate of indirect taxes applied on the value of products and services is the same for all people, whether they are poor, middle class or rich. The tax on the circulation of goods and services (TCGS) and the tax on the provision of services (TPS) are examples of indirect taxes.

- **Fees:** The shipping fee and the garbage fee are examples of fees.

- **Contributions:** the contribution to National Institute of Social Security (NISS) is an example of a contribution.

- **Regressive taxes:** the amounts of regressive taxes levied on poor people are the same amounts levied on rich people. For example, considering that the amount spent on fuel at gas stations is BRL 400.00 per month, the value of the 18% TCGS tax is BRL 87.80 per month (400.00/0.82 = 487.80 – 400.00 = 87.80), a value that is exactly the same for all people, whether they are poor, middle class or rich. This value of BRL 87.80 of TCGS per month is equivalent to 7.24422% of the income of a person who earns BRL 1,212.00 per month and is equivalent to 0.73166% of the income of a person who earns BRL 12,000.00 per month. Therefore, it is regressive because it is more costly for those who earn less and less costly for those who earn more.

- **Product cost in industry:** in industry, product cost is the sum of all variable costs and all fixed costs used to manufacture a given product.

Variable costs are formed by raw materials, whose quantities and costs vary according to the quantity produced. For example, the total cost of blender electric motors varies according to the number of blenders produced.

Fixed costs do not vary with the quantities produced. For example, the TUTP value of the blender factory shed is always the same, regardless of the number of blenders produced. The distribution (apportionment) of fixed costs per product is equal to the manufacturing cost of the product multiplied by the distribution (apportionment) factor of the fixed cost. The distribution factor (apportionment) is obtained by dividing the value of the sum of production costs of products manufactured in the current month by the sum of fixed cost values for the current month.

- **Product cost in commerce:** In commerce, the product cost is equal to the purchase cost of the product to be sold plus the fixed cost. The distribution (apportionment) of the fixed cost per product is equal to the purchase cost of each product purchased multiplied by the distribution (apportionment) factor of the fixed cost. The distribution factor (apportionment) is obtained by dividing the sum of the values of the products purchased in the current month by the sum of the fixed costs of the current month.

- **Tax payer:** only individuals who are final consumers of products and services are tax payers. Legal persons (companies) are not tax payers because they do not pay taxes, they only collect the taxes included in the price of goods and services purchased and paid for by individuals, who are the ultimate consumers of all goods and services sold.

Example:

Product cost = BRL 100.00

+ 10% profit = BRL 111.11 (100.00/0.90 = 111.11)

Product price = BRL 111.11 (before taxes)

+ 15% income tax = BRL 19.60 (111.11/0.85 = 130.71

      – 111.11 = 19.60)

+ 18% TCGS = BRL 24.39 (111.11/0.82 = 135.50 –

      111.11 = 24.39)

Product price = BRL 155.10 (for the final consumer)

When the final consumer buys the product in the example described above, he pays the legal person (company) the total amount of BRL 155.10, BRL 111.11 being the price of the product, BRL 19.60 for income tax and BRL 24.39 for

TCGS. The legal person (company) collects BRL 19.60 of income tax to the Federal Government, collects BRL 24.39 of TCGS to the State Government and keeps BRL 111.11 for her.

The current Brazilian tax system, with more than eighty taxes, is very complicated and expensive. Therefore, legal persons (companies) needed to create tax accounting, also known as tax accounting, to take care of the administration of all the procedures necessary for the collection of taxes.

Osvaldo Dalla Coletta

Economist CRE-SP n. 6.813